GRIN

Adaption to floodings with new strategies. Benue State in Nigeria as an example

Benjamin Adeagbo

Bibliographic information published by the German National Library:

The German National Library lists this publication in the National Bibliography; detailed bibliographic data are available on the Internet at http://dnb.dnb.de.

ISBN: 9783346262134
This book is also available as an ebook.

Print and binding: Books on Demand GmbH, Norderstedt, Germany
Printed on acid-free paper from responsible sources.

GRIN web shop: https://www.grin.com/document/933859

SOCIO-ECONOMIC FACTORS INFLUENCING ADAPTATION STRATEGIES OF FLOODING IN BENUE STATE, NIGERIA.

ABSTRACT

The study examined Socio-economic factors influencing adaptation strategies of flooding in Benue State, Nigeria. The study used purposive, multi-stage random, and convenient sampling techniques to select 315 farmers whose farms have been affected by flooding. Data for the study were collected from both primary and secondary sources using structured questionnaires, interviews, journals, data from NIMET and publications from other relevant agencies like BNARDA. The data collected were analysed using descriptive statistics, Maximum Likelihood Estimate (MLE).The mean age of the farmers in the study area is 47 years. The analysis of the gender indicates that farming activities are dominated by males. It was revealed that 79.0% of the household heads were married. Also, 45.7% of the farmers have primary school education, while only 28.5% (Secondary School holders) constitute lower mean averages in terms of educational qualifications. Maximum Likelihood Estimate showed that the Log Likelihood was 590.543, while Chi-Square value 12.961 was significant at 1% level of probability, this implies that the overall effect of the explanatory variables was statistically significant.

Table of content

INTRODUCTION

Economic losses caused by floods are rising in Africa(Hoeppe and Gurenko,2013). Both researchers predicted that if nothing is done by way of mitigation, crop yields would drop by 50% in 2017. This scenario is already manifesting in Asia and other tropical countries where the rural farming households depend on agriculture for their livelihood. The floods have adversely affected billions of people mostly through loss of farms and farmlands, rendering people homeless.

Floods constitute major of menace to different continents of the world at different times ranging from flash floods to pluvial floods. The notorious attitudes and events they constitute if not well managed, can lead to excessive damage. More so, these challenges are often prominent in Asia and Africa due to lack of adequate planning of roads, farmlands and other areas of critical usage which may be needed for urban and rural regions. The socio-economic factors influencing adaptation strategies of flooding play significant roles in the application of major ways of farmers plant and harvest a large chunk of harvests. They are very important and constitute how well crop production in a country thrives. It is of necessity that in the period of adverse climatic events like flooding, these factors become critical in justifying more of harvests or otherwise, the need to survive during such periods of adversity. It is noteworthy also that environmental issues or natural events like flooding can stiffen food production, therefore, the knowledge the socio-economic factors like Age, Sex, Educational Qualification and so on, will help in ascertaining the adaptive knowledge of farmers.

In Nigeria, It is very common-place event to lose a large chunk of produce to flooding based on lack of adaptive strategies hinged on socio-economic factors of farmers. Farmers often lack the meteorological knowledge of what it takes to synchronize the adaptive strategies to the climatic events of flooding. It is estimated that tons of useful Agricultural produce are washed away during excessive rainfall (flooding) and pushes to great economic losses by the nation at large and farmers in particular.

In Benue State, it is becoming increasingly difficult for farmers to engage in crop production activities during flooding without adequate adaptation strategies. These strategies are linked to Socio-economic factors and these Socio-economic factors create a lot of challenge if they are not properly understood because they present problems that will make farmers adopt the adaptive strategies slowly. The effect of these on Nigeria and theme state is lack of food sufficiency. It is very important to note that Benue State regarded as the food basket of the nation will have a major setback in fulfilling this note. The role stands out when farmers are able to churn out a harvest commensurate to the expectation of teeming population of the state and nation.

Benue State is proudly referred to as the 'food basket of the nation' since the rich nutrients deposits of alluvial soils that support bumper harvest have helped farmers in producing crops on large scale. However, with the climate change and River Benue overflowing its bank, flooding has become a critical issue in recent years. Therefore, climate change and its attendant climate events (especially floods) have become what farmers will have to cope with, since it is fast becoming unpredictable to give accurate account of crop yields on farms. Farmers therefore need adequate knowledge on the nature and causes of climate change with the attendant climatic events and the various mitigation, adaptation and coping strategies to use. This of course, depends on their access to credible information sources and their capacity to apply the information. A major problem for crop production in Nigeria (at large), and Benue State (in particular), as it concerns climate change is the reduction of arable lands which arises from the incursion of sea to arable land for farming..

Statement of the Problem

Most farmers in Nigeria depend on rain-fed agriculture and hence fundamentally are dependent on the vagaries of weather. Climatic events (especially flooding) is impacting negatively on ecosystems, farming systems and other livelihood processes. The problem of flooding impact has been significant in the reduction of crop production in Benue State with the change in cropping patterns. The change in cropping patterns has made Benue state gradually losing the acronym 'food basket of the nation' due to the lowering of agricultural output by this scourge. In recent years, there has been a decline in Benue's agricultural produce. In previous years, groundnut was produced in commercial quantity in Benue State but the situation is no longer the same in recent time. As crops decline so are food prices rising. Cultivation of the crop has gradually dropped. (Ripples, 2018)

The problem has resulted in the situation where the present crops(Yam ,Rice, Maize, Sorghum and other cereal crops) produced in large quantity to replace this commercial quantity of groundnut are expressing great flooding impact due to climate change. The scourge has affected crops both in quality and quantity. In October 2017, Nigeria's tubers of yam exported to the United States were rejected due to low quality. There is some evidence that climate change is already having a measurable effect on the quality and quantity of food produced globally. Farmers are no longer able to farm due to challenges posed by climate change. Two hectares of rice farmlands were washed away by heavy rainfall in Benue State as a result of climate change in August 2017. Over 3,000 other farmlands were also submerged, affecting about two million farmers in the state. The overflow of water from the River Benue coupled with the excess water from River Niger have increased the incessant occurrence of flooding in Benue State (Ripples, 2018).The severity of flooding was also noted in the months of September and October, 2012, when it ravaged some parts of Nigeria especially Benue State and the quantification of such economic loses on crop production was quite immense. Cereal crops were the worst hit with over 75% of crops like rice, maize, sorghum suffering economic loses and 50% of yams cultivated had specific quality challenges. These challenges are products of unprecedented flooding which destroyed hundreds hectares of farmlands and damaged crops as a result of different heavy downpour, river overflows (especially River Benue) among others. (Adeloye and Rustum , 2014).

Flood occurrences are fast becoming annual events which are largely due to climate change. These occurrences are common during rainy season and basically hovered between May and October of every year. One of such years is . The event pushed most of the country's rivers over their banks and submerged hundreds of kilometers of urban and rural land. Overall, an estimated 1.3 million people were displaced and about 431 people lost their lives with several hectares of farmland destroyed. Though the unusually large flood was predicted by the Nigeria Metrological Agency NIMET, government at all tiers failed to act on time, resulting in the worst humanitarian crises in Nigeria since the civil war in 1967-1970. Despite the expected increase in frequency and magnitude of flood in Nigeria and invariably Benue, few economic impact assessment studies on crop production have been undertaken to establish the underlying causes of their vulnerability. In the absence of comprehensive data and information, the measures to cope with flood have remained ad hoc. Agricultural sector is the most affected sector of Benue economy. This is because most of the economic activities around or within flood prone areas are agriculture-oriented and 75% of the livelihood activities in agricultural sector. In crop production, farmers lost 21.7billion Naira to flood disaster which adversely affected the rural economy with negative multiplier effect on Benue State economy.

Objective of the Study

The objective of the study is to determine the Socio-economic factors influencing adaptation strategies of flooding in Benue State, Nigeria.

METHODOLOGY

The Study Area

The study was carried out in Benue State. Benue state is one of the North central states in Nigeria. The state has 23 local government areas namely: Buruku, Gboko, Guma, Gwer east, Gwer west, Markudi, Tarka, Ado, Agatu, Apa, Obi, Ogbadigbo, Ohimini, Oju, Okpokwu, Ukum, Katsina-ala, Konshisha, Kwande, Logo, Otukpo, Ushongo, Vandeikya respectively.

Benue state lies within the lower Rivers Benue trough in the middle belt region of Nigeria. Its geographic coordinates are longitude $7^{0}47'$ and $10^{0}0'$East. Latitude $6^{0}25'$ and $8^{0}8'$ North and shares boundaries with five other states namely: Nasarawa State to the north; Taraba state to the east, cross river state to the south, Enugu state to the south west and Kogi state to the west. The state also shares a common boundary with the republic of Cameroon on the south-east. It also occupies a land mass of 34, 059 square kilometers (LCAPC (2014))

Agriculture forms the backbone of the Benue State economy, engaging more than 70 percent of the working population. This has made Benue the major source of food production in the Nation. Mechanization and plantation agriculture/agro-forestry are still at its infancy. Farm inputs such as fertilizers, improved seed, insecticides and other little foreign methods of agro-chemicals are being used. However, cost and availability is still a challenge. Important cash crops include Soya-beans, Rice, Peanuts, mango varieties, Citrus. Other cash crops include Palm oil, Melon, African pear, Chili pepper, Tomatoes etc. food crops include Yam, Cassava, Sweet potato, Beans Maize, Millets,

Guinea corn, Vegetables etc. there is very little irrigation agricultural techniques. Animal production includes Cattle, Pork, Poultry and Goat but no dairy products.

Sampling Technique and Sample Size:

The state was divided into three Agricultural zones: A, B and C accordingly;

Zone A ; Benue North Central: Buruku, Gboko, Guma, Gwer east, Gwer West, Markudi, Tarka.

Zone B: Benue South : Ado, Agatu, Apa, Obi, Ogbadibo, Ohimini, Oju, Okpokwu

Zone C: Benue North-east: Ukum, Katsina-Ala, Konshisha, Kwande, Logo, Otukpo, Ushongo, Vandeikya.

The population for the study consists of farmers whose farms were affected by flooding in the study area. Purposive and multistage sampling were adopted for the study. The First stage involved the purposive selection of four Local Government Areas (4 LGAS each) from the three Agricultural Zones while Second stage was the purposive selection of the farmers in the twelve (12) Local Government Areas through the Agricultural Development Program (ADP) enumerators. Thirdly, (3) extension blocks were selected from the 12 LGAS, making thirty-six (36) extension blocks in all. Fourthly, one (1) farming village/community was selected from each extension block making a total of (12) villages/communities. Lastly, through convenient sampling and findings of (less-prone to flooding areas (zone A) to more-prone to flooding areas(zones B and C), questionnaires were administered.

Number of Farmers affected by Flooding(Sampling frame) and the number of Farmers(Sample Size) in the Local Government Areas and Villages(Communities)

Agricultural Zones	LGA	Extension Blocks	Villages/Co mmunities	Distribut ion Of Question naires	Total	Sample size
Zone A	Markudi	3	1	30	100	83
	Gboko	3	1	15		
	Gwer-East	3	1	25		
	Gwer-West	3	1	30		
Zone B	Ado	3	1	40	125	107
	Agatu	3	1	25		
	Apa	3	1	25		
	Ogbadibo	3	1	35		
Zone C	Katsina-Ala	3	1	25	135	125
	Konshisha	3	1	30		
	Otukpo	3	1	50		
	Vandeikya	3	1	30		
Total	12 LGAS	36 Extension Blocks	12 Village/Co mmunities		360	315

Method of Data Collection

Data for the study were collected from primary and secondary sources respectively. Primary data were collected through questionnaires with the help of Agricultural Development Programs enumerators of the selected state, at the extension block level. Primary data were also collected using structured interview schedules. The data collection instrument focused on capital resources used, agrochemicals used and other relevant information. .The secondary data were collected from Ministries, Agricultural and other relevant agencies, like NIMET (Nigeria Meteorological Agency), Benue State Meteorological Agency and Benue Agricultural and Rural Development Agency (BNARDA).

Method of Data Analysis

The following analytical tools were used to achieve the objective in the study:

i. Descriptive Statistics
ii. Maximum Likelihood Estimation (MLE) Method

RESULTS AND DISCUSSIONS

Socio-Economic Characteristics of the Respondents (Farmers).

The Socio-economic characteristics of respondents (farmers) affected by flooding is presented in Table below. The result shows that 71.4% of the farmers were males, while 28.6% were females. This implies that farming in the study area is dominated by male and it is also of significance that more female will have to be encouraged to participate in farming in the study area. Also, it is shown that 73.8% of the farmers fall within the age bracket of 42-51, while 26.2% fall into 52-61 years. This means that majority of the farmers are within the working age group. The study also revealed that 79.0% of the household heads were married and this agrees with Ziervogel (2006) who agreed that men who were married have easier access to farmland through paternal inheritance.

Education plays a significant role in decision making skill acquisition and enhancement of one's ability to understand to plan and to plan and take risks. In Table below, 45.7% of the farmers have primary school education. This has implication especially as it concerns adaptation strategies in the study area as adoption of adaptation strategies will come with slow pace due to this percentage. Only 28.5% (Secondary School holders) and 25.7% (Tertiary School holders) constitute the lower mean averages in terms of educational qualification. The household size in the study area shows, 20.6% of the household size in the study area constitutes (1-5) number of persons per household, while 52.1% of the household size constitutes (6-9) number of persons per household. This result is lower than the average of 9 persons per household as reported by (Irohibe and Agwu, 2014) in Kano State, Nigeria that large household size could play a great role in family labor provision in the agricultural sector, but could place greater burden on non-farming households.16.5% constitutes (10-13) number of persons per households, 4.76% constitutes (14-17) number of persons per household, 2.54% constitutes(18-21) number of persons per household, while 1.90% and 0.63% constitutes (19-25, and 26-29) number of persons per household respectively. The result also shows 79.1% of the farmers take farming as their main occupation or means of livelihood while 18.3% civil servants take farming as a part-time means of livelihood. More so, 2.6% students use farming to support themselves. On accessibility to credit, only 9.52% of the farmers, that is 30 farmers, have access to credit, while, 90.48% (285) of them have no access to credit facilities. Estimated monthly income showed a greater proportion (65.08%) of the household heads realized between ₦40,000 and ₦ 60,000 per month, which means most of household heads still operate on a low income per farm basis.

Socio-Economic Characteristics of the Farmers

Variables	Frequency	Percentage
Gender		
Male	225	71.4
Female	90	28.6
Age		
22-31	2	5.3

32-41	52	15.5
42-51	141	43.8
52-61	88	26.2
62-71	32	9.2
Marital Status		
Married	249	79.0
Single	28	8.9
Widows	28	8.9
Divorced	10	3.2
Educational Qualifications		
Primary School	144	45.7
Secondary School	90	28.5
Tertiary School	81	25.7
Household Size		
1-5	65	20.6
6-9	164	52.1
10-13	52	16.5
14-17	15	4.76
18-21	08	2.54
19-25	06	1.90
26-29	02	0.63
30 and above	03	0.95
Means of Livelihood		
Farmers (main occupation)	249	79.04
Civil Servant (Part-time occupation	58	18.41

Students (support occupation)	8	2.54
Access to credit		
Yes	30	9.52
No	285	90.48
Estimated Monthly Income(Naira)		
Income ≤ 20,000	17	5.40
20,001 ≤ 40,000	33	10.48
40,001 ≤ 60,000	205	65.08
60,001 ≤ 80,000	47	14.92
80,001 ≤ 100,000	07	2.22
100,001 and above	06	1.90
Total	**315**	**100**

Source: Data Computed 2019

Maximum Likelihood Estimates (MLE) of the Ordered Logit Model presented in Table 4.2 shows that out of the nine explanatory variables included in the Ordinal regression model, the coefficient of Educational Status (-0,063), Occupation (-0.221) and Household weekly spending (6.896) were the significant factors influencing adaptation strategies among the respondents.

Education is negative and significant at 1% level of significance. This indicates that one unit decrease in level of education would result in a 0.063 unit increase in the ordered log-odds of being in a higher socio-economic status category while the other variables in the model are held constant.

Occupation is negative and significant at 1% level of significance. This indicates that one unit decrease in occupation would result in a 0.221 unit increase in the ordered log-odds of being in a higher socio-economic status category while the other variables in the model are held constant. Thus, for a one unit increase in the occupation, the odds of high socio-economic status versus the combined middle and low socio-economic status categories are 0.221 times greater, given the other variables are held constant in the model.

Household weekly spending is positive and significant at 1% level of significance. This indicates that one unit increase in household weekly spending would result in a 6.896 unit increase in the ordered log-odds of being in a higher socio-economic status category while the other variables in the model are held constant. Thus, for a one unit increase in the household weekly spending, the odds of high socio-economic status versus the combined middle and low socio-economic status categories are 6.896 times greater, given the other variables are held constant in the model. Likewise, for a unit decrease in household weekly spending, the odds of the combined high and middle socio-economic status versus low socio-economic status are 6.896 times greater, given the othesr variables are constant.

Maximum Likelihood Estimate shows that the Log Likelihood was 590.543, while Chi-Square value 12.961 was significant at 1% level of probability, this implies that the overall effect of the explanatory variables was statistically significant.

Maximum Likelihood Estimate (MLE) of the Ordered Logit Model

Variables	Coefficient	Error	Wald	Sig.	Dec.
Age(X_1)	0.011	0.011	1.022	0.312	Not Sig.
Educational Status	-0.063	0.023	7.356	0.007***	Sig
Occupation(X_2)	-0.221	0.039	16.521	0.000***	Sig
Household spending (X_4)	6.896	2.001	0.457	0.001***	Sig
Sex (X_5)	0.331	0.235	1.935	0.164	Not Sig

-2 Log Likelihood for the Model: 590.543

No. of Observation 315

Chi-Square: 12.961***

Prob> Chi-Square : 0.0000

*Sig= Significant, ***-Significant at P≤0.0, Dec, =Decision*

SUMMARY, CONCLUSION AND RECOMMENDATIONS

Summary of the Findings

The study dwelt on Socio-Economic factors influencing Adaptation strategies of flooding in Benue State, Nigeria. Cross sectional and time series data were used. Also, the primary data were collected using questionnaires. Interview schedules were also used. The questionnaires were administered to 360 respondents (farmers) using multi stage sampling method. 315 questionnaires were returned and worked on.

The data were analyzed using descriptive statistics, Maximum Likelihood Estimation on farming household data that were spread within the 23 Local Government Areas of Benue State and divided into three Agricultural zones. These were Zone A: include Makurdi, Gboko, Gwer-Eat and Giver West, Zone B; Ado, Agatu, Apa and Ogbadigbo and Zone C :KatsinaAla, Konshisha, Otukpo and Vandeikya respectively. The results of the socio-economic characteristics show that a greater proportion (71%) of the farmers affected by flooding were males while (73.8%) of these farmers fall between the age group of 42 and 51 years. 79.0% of farmers were married and 45.7% of the farmers have primary education. Furthermore, 52.1% fall (6-9) number of households, while 79.1% of farmers earned their living from farming.

Conclusion

Socio-Economic factors played a significant role in adaptation strategies apply to planting crops during flooding in the study area. This means flooding is a very important parameter and having adverse effects on crop production in Benue state, therefore, the need for adaptation strategies in the area. More so, farmers engaging in adaptation strategies in the area have opportunities to stay afloat during flooding despite constraints faced by them in the employment of these strategies. The decrease in crop yields is a threat to crop production in the state in particular and Nigeria in general.

Recommendations

1. There is need for the establishment of Meteorological station units in rural farming area especially in Benue State where accessibility is extremely difficult to make available

meteorological data (information) on climatic variables to the reach of farmers in the rural areas.

2. A large number of farmers in Benue State are poor and adaptive capacity is limited. Therefore, there is need to provide relief services for them to be able to adapt to the impacts of flooding. Furthermore, Agricultural Extension Service will play a significant part in showing the farmers on how best to adapt to flooding impacts.

3. Since socio-economic factors influence or drive adaptation strategies in Benue State, therefore, it will be important that the Government policies should be aimed at raising the economic status of farmers such as level of education of the household heads and improving access to education for farmers. This will help in empowering the labour force on farms and reduce the dependency rate. Also, it will be necessary to raise the low weekly spending of farmers by providing sustainable loans through the establishment of Agricultural-based micro-finance banks that will be charged with the provision of these loans. In addition, Farmers should be encouraged to join co-operative societies and organise one in area where none exists since these co-operative societies often help in raising farmers' income.

4. Each of the constraints to choice of adaptation strategies faced by farmers in Benue State has significant threat to crop production and it is important that these constraints are specifically dealt with based on their peculiarities. On lack of proximity of adaptation strategies to farm areas, it is recommended that the adaptation strategies should be close enough to farm areas to avoid this challenge which often discourage farmers from adopting such strategies. The training of farmers by Extension Agents is also important to ease-off the usage of both some strategies and tillage practices that help in mitigating flooding in the area.
5. Socio-Economic factors constitute a major influence that farmers used in ameliorating menace of flooding and must obtain a thorough look for policy makers.

References

Adeloye, A. J. and Rustum, R. (2011). Lagos (Nigeria) flooding effects *Urban Design and Planning,* 170(DP4): 180-85.

Adeloye, A. J. and Rustum, R. (2014). Lagos (Nigeria) flooding and influence of urban planning. *Urban Design and Planning,* 164(DP3): 175-87.

Hoeppe, P. and Gurenko, E. N. (2013). Scientific and economic rationales for innovative climate insurance solutions. *Climate Policy,* 6(6): 607-20.

Irohibe, I. J. and Agwu, E. A. (2014). Assessment of food security situation among farming households in rural areas of kano state, nigeria. *Journal of Central European Agriculture,* 15(1): 94-107.

Ripples (2018). Benue and consequences of flooding 180 (DPR) 195-210.

Appendix

```
PLUM LADAPT BY SEX WITH AGE EDU OCCUP HHSPEND
  /CRITERIA=CIN(95) DELTA(0) LCONVERGE(0) MXITER(100) MXSTEP(5) PCONVERGE(1.0E-
6) SINGULAR(1.0E-8)
  /LINK=LOGIT
  /PRINT=FIT PARAMETER SUMMARY

  /SAVE=ESTPROB.
```

PLUM - Ordinal Regression

[DataSet0]

Warnings

There are 560 (68.0%) cells (i.e., dependent variable levels by combinations of predictor variable values) with zero frequencies.

Case Processing Summary

		N	Marginal Percentage
LADAPT	1	96	30.5%
	2	146	46.3%
	3	65	20.6%
	4	8	2.5%
SEX	0	90	28.6%
	1	225	71.4%
Valid		315	100.0%
Missing		0	
Total		315	

Model Fitting Information

Model	-2 Log Likelihood	Chi-Square	df	Sig.
Intercept Only	603.504			
Final	590.543	12.961	5	.014

Link function: Logit.

Goodness-of-Fit

	Chi-Square	df	Sig.
Pearson	782.237	610	.000
Deviance	501.630	610	.999

Link function: Logit.

Pseudo R-Square

Cox and Snell	.040
Nagelkerke	.045
McFadden	.018

Link function: Logit.

Parameter Estimates

		Estimate	Std. Error	Wald	df	Sig.	95% Confidence Interval	
							Lower Bound	Upper Bound
Threshold	[LADAPT = 1]	-1.224	.613	3.994	1	.046	-2.425	-.024
	[LADAPT = 2]	.870	.610	2.032	1	.154	-.326	2.066
	[LADAPT = 3]	3.336	.694	23.102	1	.000	1.976	4.697
Location	AGE	.011	.011	1.022	1	.312	-.010	.032
	EDU	-.063	.023	7.358	1	.007	-.109	-.018
	OCCUP	-.221	.039	16.521	1	.000	-.494	.052
	HHSPEND	6.896	2.001	.457	1	.001	-1.196	2.455
	[SEX=0]	.331	.238	1.935	1	.164	-.135	.798
	[SEX=1]	0[a]	.	.	0	.	.	.

Link function: Logit.

a. This parameter is set to zero because it is redundant.